The CHEMISTRY of Cosmetics

Dear Reader

The global cosmetics industry is worth many billions of dollars, and creates and sells thousands of products to people worldwide. Many consumers today are interested in how cosmetics are manufactured and tested, and they read the ingredient lists on product labels. This book contains information about the ingredients and basic chemistry involved when companies produce cosmetics, including everyday products like toothpaste and sunscreen.

In Chapter 5, find out how public pressure has brought about bans on animal testing in cosmetic production. In part, these restrictions were made possible when enough people learnt more about the chemistry of cosmetics.

> "ROSEHIP OIL WAS LIKE NO OTHER COSMETIC PRODUCT ..."
>
> FOUNDERS OF TRILOGY

I hope that you will be inspired to do more of your own research on how cosmetic products are made, so you can make informed choices when you buy them.

Sharon Parsons

My sincere thanks to the following people for their time, information, images and enthusiasm for this book:

The team at Trilogy, Wellington, New Zealand

Chris Ryan, Silverson Machines Ltd, Chesham Bucks, the UK

For learning solutions, visit cengage.com.au

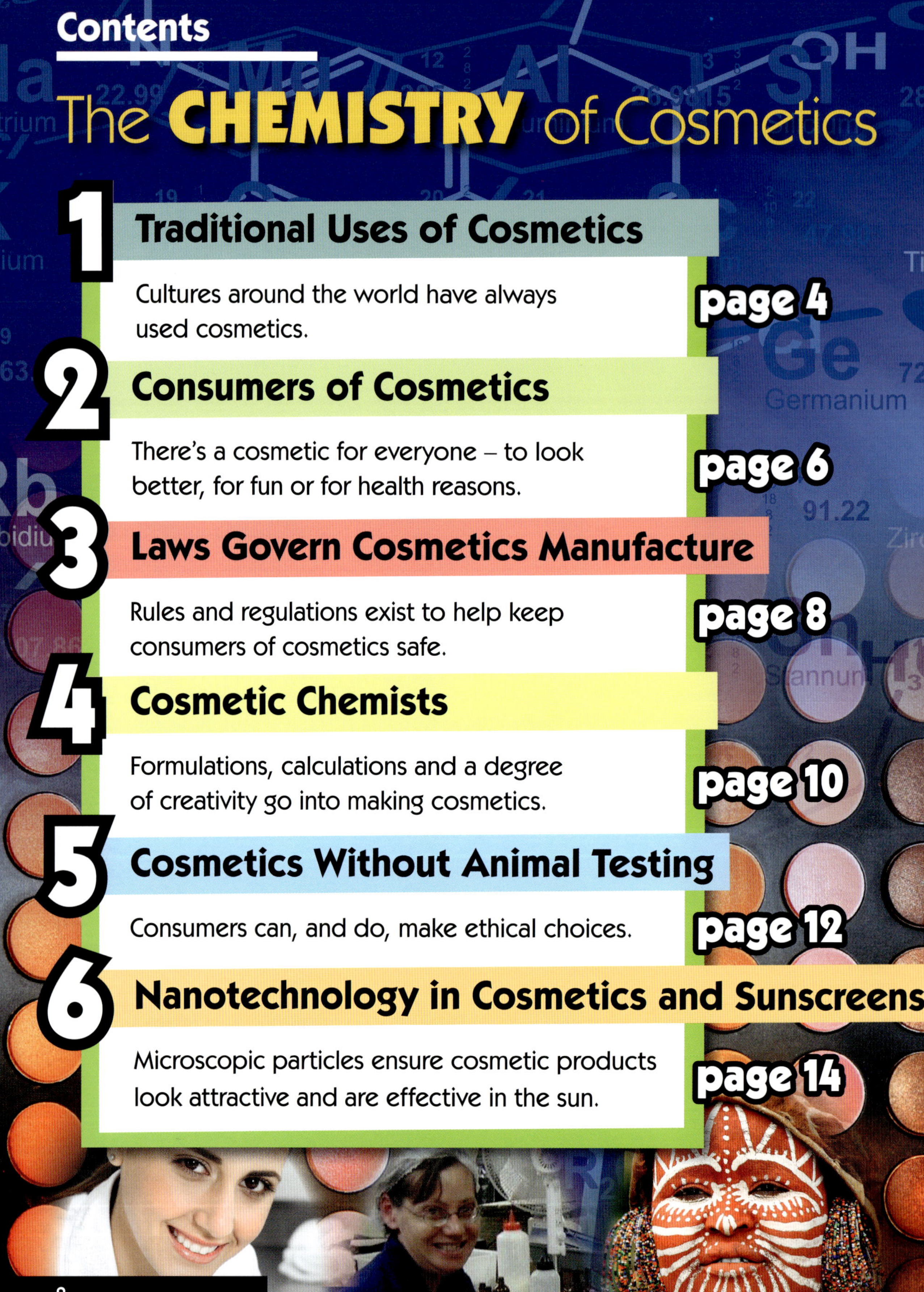

Contents

The CHEMISTRY of Cosmetics

1 Traditional Uses of Cosmetics

Since the earliest civilisations, in almost every society around the world, people have used cosmetics to enhance their appearance. In fact, the word "cosmetic" comes from the Ancient Greek word *kosmetike*, which means "the art of dress and ornament".

South America, Australia and the Pacific Islands

Indigenous peoples in places such as South America, Australia and the Pacific Islands have long used paints and other facial or body decoration for ceremonial events, for battle or to enhance the performance of traditional stories and dance.

In South America, an Amazonian Indian woman uses red paint to dye areas around her eyes and nose.

In Papua New Guinea, a Pacific Island country, a woman's make-up prepares her for the skeleton dance in a tribal festival.

In Australia, a member of the Nunukul Yuggera dance group portrays the Dreaming story of the goanna.

Africa and India

In Africa and India, many people used natural substances, such as henna, to dye their hair and skin. They used beeswax mixed with the sap of certain plants to make hair gel.

henna decoration on the hands of a bride in India

A mother paints her child's face according to tradition in a village near Jodhpur, India.

an African woman (left) and man (right) with traditionally painted faces

China

In early Chinese society, people used colours on their fingernails as a mark of their social status. Royalty wore gold and silver, while workers were forbidden to wear any bright colours.

an ancient Chinese make-up cosmetics box

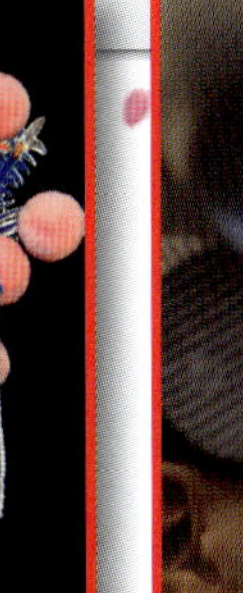

a Shanxi Opera character in full make-up and costume

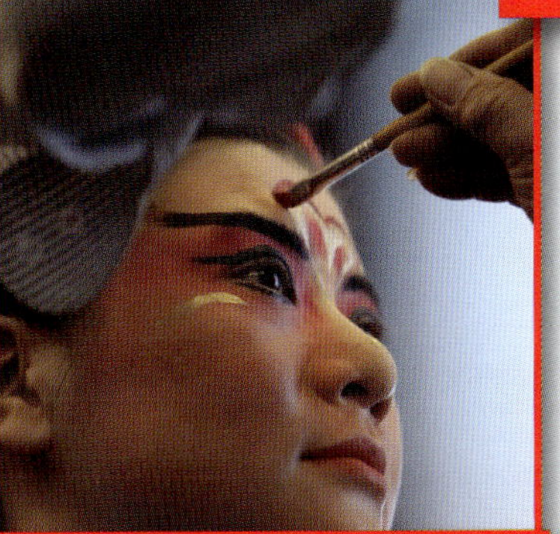

An operatic actress in Beijing has a scenic mask painted on her face.

Nepal

In Nepal, festivals are an important part of the culture. At many festivals, it is common for babies and small children to wear make-up using rich shades of red and yellow, and thick black lines around their eyes.

Japan

Since the 1600s in Japan, performers in traditional kabuki theatre have worn make-up called *kushi*. This make-up is integral to the story being dramatised. All performers apply a thick, white face paint before adding the detailed features.

an actor dressed to perform in Japanese Kabuki theatre

a Nepali boy in make-up for a traditional festival in Kathmandu, Nepal

England's Queen Victoria (1819-1901) thought that cosmetics were improper, vulgar and should only be used by actors.

Europe

In wealthy or aristocratic European societies, it was popular among both men and women to wear cosmetics that made their skin appear whiter, as a sign of their status. Unfortunately, the most common face-whitener was white lead paint, mixed with arsenic, both of which were extremely poisonous. The search for beauty and status through the use of cosmetics sometimes proved harmful.

2 Consumers of Cosmetics

In today's society, people have become accustomed to an array of cosmetics and body-care products to choose from. The most popular are used on the face, hair and nails. Consumers, or people who buy the products, can have an insatiable appetite for cosmetics and body-care products, to improve or enhance their skin and appearance. This has fuelled the emergence of many large and powerful cosmetic companies to create a worldwide industry, which today, is worth over $US170 billion* each year. In Australia, sales of cosmetics and toiletries amount to about five billion dollars each year. (**Source: French market research firm Eurostaf, 2007*)

Cosmetic Categories

Make-Up, Just for Fun

While most people apply make-up for everyday use, it has other uses, too, such as face painting for occasions like stage performances, role plays, sports events or just for fun.

role-play make-up

clown make-up

make-up for Halloween

make-up to support sports teams

Make-Believe Make-Up

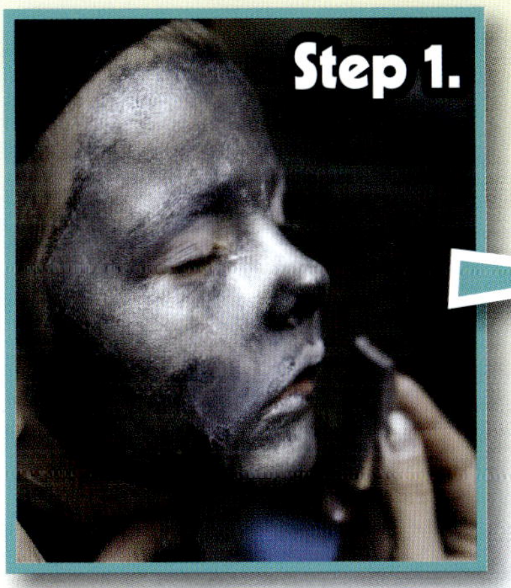

Step 1.

Step 2.

Step 3.

Step 4.

Step 5.

ROAR!

3 Laws Govern Cosmetics Manufacture

As a corporate lawyer, I write laws and regulations for cosmetic manufacturing companies to follow.

Cosmetics and personal care products are specially formulated and manufactured under strict government regulations to ensure that they are safe for people to use.

The Australian government classifies the ingredients in cosmetic products as industrial chemicals. As such, the government regularly reviews the regulations and conditions by which cosmetics can be manufactured and imported, so that consumer health is protected. Government regulations also protect workers handling the raw ingredients in cosmetic manufacturing companies.

inside a cosmetics manufacturing facility

an automated assembly line in a cosmetics factory

Packaging Labels

Government regulations extend to cosmetic packaging. Companies are required to disclose a complete list of ingredients so that consumers can make informed choices that suit their needs and health requirements.

Therapeutic Cosmetics

Therapeutic cosmetics are standard products that have had ingredients added to protect or heal the body. An example of a therapeutic cosmetic is a skin moisturiser with extra ingredients added, to give people's skin some protection from the sun's UV rays.

Category	Product Examples
Face and Nail	• skin foundations with sunscreen • nail polish with fungal infection prevention
Skin Care	• moisturisers with sunscreens • antibacterial skin products • anti-acne products
Oral Hygiene	• products to treat teeth and mouth
Hair Care	• anti-dandruff products

Non-Therapeutic Cosmetics

Non-therapeutic cosmetics are made to beautify the body or perform a body-care function. They are not formulated to protect or heal the body.

Category	Product Examples
Face and Nail	• eye shadows, lipsticks, nail polishes
Skin Care	• moisturisers
Oral Hygiene	• toothpaste, tooth gels
Hair Care	• hair colour products, shampoos, conditioners
Perfumes	• perfumes, colognes
Personal Hygiene	• deodorants, bath gels, soaps

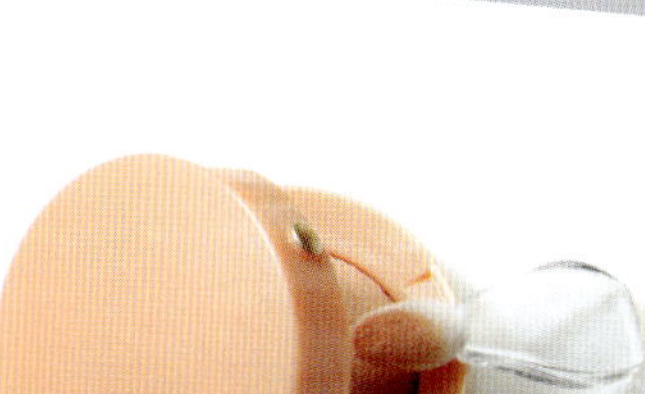

4 Cosmetic Chemists

The lessons learned about health and safety in cosmetics manufacturing have led to allied industries and jobs, such as cosmetic testing laboratories and cosmetic chemists. Most cosmetic chemists have a university degree in one of these areas: chemistry, chemical engineering, biology or microbiology. They may also have an advanced degree in cosmetic science.

cosmetic chemists working as a team

Cosmetic chemists usually have a particular speciality. A synthesis chemist creates the raw materials for cosmetics. Chemical synthesis is the process of combining simple elements to create more complicated, artificial substances. Most synthesis chemists work for companies that specialise in supplying raw materials.

an extract from the aloe plant used in some natural cosmetics

Formulation Chemists

A formulation chemist uses the materials produced by synthesis chemists to create formulas for new cosmetic and body-care products. Formulation chemists have to be creative to come up with new products with useful properties and special features. Most formulation chemists work for cosmetics manufacturers.

Product Specialisation

Formulation chemists must have a thorough knowledge of the raw ingredients used in cosmetic manufacturing. They may create formulas for many different kinds of products, including make-up, perfumes, and toiletries such as deodorants or soaps. But they usually begin their career by specialising in one kind of product, such as shampoo.

Quality Control Chemists

New formulas are handed over to a quality control chemist for a long testing and trialling process so that the final formulation can be produced for consumers to use. Quality control chemists check every product to make sure it is meeting government regulations.

Improving Formulations

A formulation chemist will produce several batches of a product formulation to test how it changes or improves when the quantities of one or more ingredients are adjusted up or down. They will also want to test what happens when they change certain conditions, such as the processing temperature or mixing time. By having a rigorous testing process, a formulation chemist can feel confident that they have arrived at a combination of ingredients and method that will yield the best product.

During the testing process, formulation chemists will also identify how to make cost savings. Formulation spreadsheets for each product list all the ingredients, quantities and prices for raw materials. As changes to the formulation are entered into the spreadsheet, the program can automatically recalculate data and costs for the product.

"This is hair wash number 1 499 – one more to go!"

Science

A Chemist Tests His Hair Formulation

One formulation chemist described his early experiences of trying to create a better shampoo and conditioner formulation than competitors' products, by testing the formulations himself. After one year of research and tests, he calculated that he had washed his hair 1500 times!

Social Studies

Australian Society of Cosmetic Chemists

The Australian Society of Cosmetic Chemists is a professional organisation of people and companies that work together to share knowledge about the science and technology involved in formulating cosmetics. As a group working in the fields of cosmetics, toiletries and perfumery, their aim is to also find ways to improve the quality and safety of the products.

5 Cosmetics Without Animal Testing

The cosmetics industry is very competitive, so every cosmetic company must create or modify products that will appeal to the changing needs and desires of their customers. For example, when certain consumer groups chose not to use cosmetics that were tested on animals, many cosmetic companies responded by adopting alternative ways to test their product ingredients. In turn, they communicated their change in testing methods to consumers via marketing campaigns. These included releasing advertisements and adding logos and slogans to their packaging with statements such as, "Not Tested on Animals."

NOT TESTED ON ANIMALS

However, in many countries, the government stipulates that certain ingredients and formulations must be tested on animals before they are deemed safe for humans to use. Many companies have simply chosen not to use those kinds of ingredients in their products.

Bubbles bubble and burst!

Buy your bottle today!

NOT TESTED ON ANIMALS

"Hey kids, this bubble bath product was not tested on animals! I love it!"

Europe Bans Testing Cosmetics on Animals

The European Commission is an organisation representing the 27 countries that belong to the European Union (EU). In 2004, the European Commission announced a ban on using animals to test finished cosmetic products.

THE EUROPEAN UNION (EU)

The EU is a partnership between 27 European countries with close political and economic relationships. With a few exceptions, they also share a currency, the euro.

The EU expanded the ban in 2009 to prevent animal testing on cosmetic ingredients as well as finished products. In 2013, the ban was officially enforced throughout the EU.

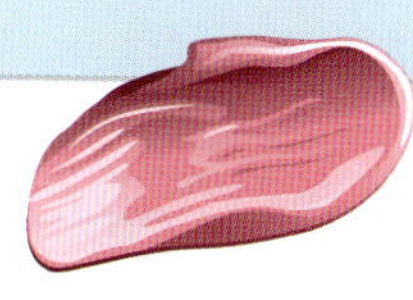

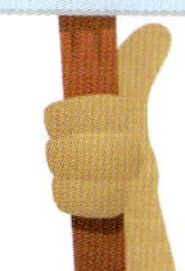

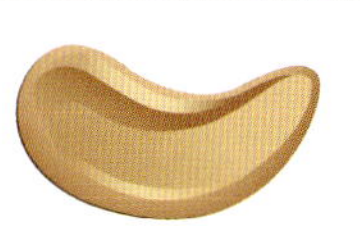

European Partnership to Alternative Approaches to Animal Testing (EPAA)

As a means of supporting the ban, the European Commission established the EPAA to provide cosmetic companies with viable alternative and humane forms of testing cosmetics.

International Cooperation on Cosmetics Regulation (ICCR)

The ban enforced by the European Commission has sent a strong message to other countries to also find alternative methods of testing ingredients and products for cosmetic use. And in 2007, regulatory bodies from a number of countries formed the International Cooperation on Cosmetics Regulation group. The group regulates cosmetics from the EU as well as Japan, Canada and the United States. One of its primary concerns is developing scientifically sound recommendations for alternative testing methods.

map of the EU

6 Nanotechnology in Cosmetics and Sunscreens

Nanotechnology is the manufacture and manipulation of organic or non-organic materials at the level of atoms. Nanoparticles are tiny particles measured in nanometres. One nanometre equals one billionth of a metre. Nanoparticles can be made up of many different elements, but they are all smaller than 100 nanometres. Now that's microscopic!

Nanoparticles in Our Environment

Nanoparticles are not new in our environment. They are present in dust, motor vehicle emissions and airborne salt particles that rise from the sea, as well as many other places.

What is new is our ability to use nanotechnology to alter the structure of organic and inorganic materials into tiny nanoparticles for all kinds of uses and products.

Nanoparticles are present in the airborne salt particles of sea spray.

Nanoparticles are present in the fine carbon emissions from motor vehicles.

Are Nanoparticles Safe in Sunscreen?

The use of nanoparticles in the formulation of cosmetics and sunscreens has raised much discussion about their possible health risk to people. In the case of sunscreens, cosmetic chemists suggest that the fine nanoparticles help companies to produce a smoother product that not only provides better coverage but also increased UV (ultraviolet) protection.

The issue of concern to the public and various consumer health groups is that the minute particles may be so fine that they will be absorbed through the skin and into the bloodstream, and adversely affect the body's cells and organs.

In 2011, a report written for the Australian Society of Cosmetic Chemists stated that there might be unknown health problems for people applying products that use nanoparticles. The report also raised health concerns for people handling such fine particles in cosmetic manufacturing companies, who may breathe in the nanoparticles and suffer chronic respiratory conditions as a result.

It is wise to apply sunscreen before and during time spent in the sun.

Natural Sunscreen

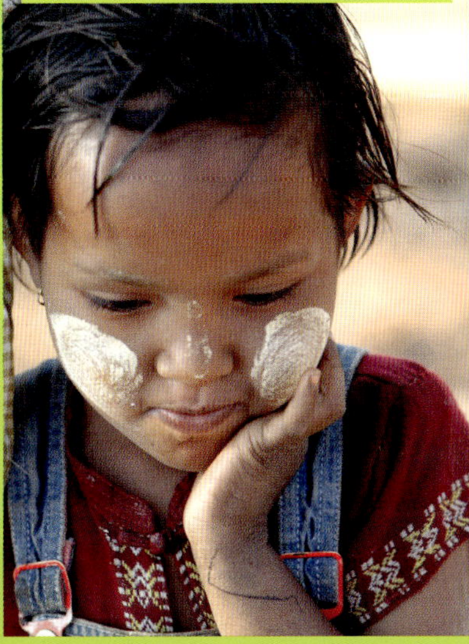

a young Burmese girl with traditional sunscreen on her face – a powdery paste made from tree bark

an image set up to compare a woman applying protective sunscreen (left) and a woman (right) who increases her risk of skin cancer by not wearing sunscreen

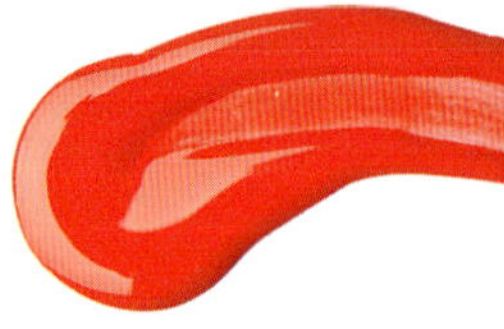

7 The Chemistry of Cosmetic Colours

Inside a cosmetic chemist's laboratory, achieving the right colour is as important as producing the right product formulation. Colours will vary from season to season according to fashion trends. So, cosmetic chemists need to have a good understanding of how to create particular shades of colours, to match fashionable season colours.

The Colour Index

Cosmetic chemists around the world refer to the international Colour Index (CI) for a complete list of the thousands of colours and their hues (or shades), as well as their technical details. Every colour has a CI number so people from different industries can find out how the CI colours are created. The international Colour Index provides information about all kinds of products and their colour composition.

POPULAR RED

Red in its various shades has throughout history been the most popular colour for lipsticks and rouge.

a small sample of the thousands of CI colours

a cosmetic chemist at work

Understanding Colour Additives

Cosmetic chemists first learn that there are two main kinds of additives that create colour in cosmetics: dyes and pigments.

a range of powder dyes

Dyes – these additives can dissolve in a formulation, or liquid.
Pigments – these additives cannot dissolve in a formulation. They are used in most cosmetics, such as lipsticks.
White Pigments – these are commonly used in the formulation of cosmetic products when good skin coverage is necessary, such as liquid foundations and eye shadow. Cosmetic chemists also use white pigments to keep products stable (retain the right texture) in warm conditions and when exposed to light.

Iron Oxide Colours

There are three core iron oxide pigments: black, red and yellow. Cosmetic chemists blend combinations of these three colours, in their different shades, to create natural colours that match people's skin tones. They are used in products like foundation, eye shadow and mascara.

IRON OXIDISES TO RED

In ancient times, iron oxides from the ground were used to create cosmetics for the face and body. These are compounds that include both iron and oxygen. When iron is exposed to weather, it oxidises, or combines with oxygen, to produce warm red, yellow and orange colours, as well as brown and black.

8 Should Potentially Harmful Chemicals be Banned from Cosmetics?

TEXT TYPE Discussion

In the past four decades, a growing number of consumers and doctors have expressed concern about the use of toxic chemicals in body-care products and cosmetics. Since the 1970s, this concern among consumers has triggered the formation of new companies specialising in body-care products and cosmetics that only use natural or organic ingredients. As scientific evidence casts doubt on the safety of using certain synthetic (unnatural) ingredients in cosmetics, more people are switching to natural products. This is a growing worldwide trend as more consumers are becoming better informed. The question is, should these toxic chemicals be banned altogether from cosmetics?

A Medical Researcher

In my scientific studies of hundreds of cosmetics, I have uncovered traces of heavy metals such as lead, mercury and cadmium. When people use cosmetics containing traces of these heavy metals, they can be absorbed through the skin and pass into the bloodstream. Medical researchers have found sufficient evidence that these heavy metals are harmful to humans. Their presence can lead to chronic health conditions such as developmental issues, dysfunctions of the immune system and cancer.

"I strongly urge government health departments to introduce legislation to have these toxic chemicals banned ..."

Serious health problems can also occur among the people who work in the cosmetic industry. There are no strict government guidelines as to what are acceptable levels of exposure to heavy metals in the workplace. Heavy metals can accumulate in the human body over time, and it is this build-up of harmful heavy metals in the body which causes me the most concern.

Many of the chemicals commonly used in cosmetics and body-care products are also harmful to the environment. Cosmetics and toiletries are usually washed down the drain after use. From there they can pass into our rivers and other waterways.

I strongly urge government health departments to introduce legislation to have these toxic chemicals banned from cosmetics and body-care products.

A Natural Cosmetic Company Director

My company, Nature and Nurture Cosmetics, is proof that it isn't necessary to use harmful ingredients to produce great cosmetics. We develop and sell products that use only natural or organic ingredients. Our customers feel confident that our products are safe to use, as they do not include artificial preservative and additives. Just as importantly, our cosmetics and body-care products create the effects that our customers want! Our array of gorgeous natural lipsticks, colourful eye shadows and organic moisturisers show that toxic chemicals simply aren't necessary.

After 20 years in business, my company has an annual turnover of 50 million dollars. We are so successful that three large companies that produce traditionally made cosmetics have offered to buy my company. This interest among established cosmetic companies is proof that enough consumers are purchasing natural cosmetics to affect the sales of traditional cosmetic products.

"I strongly recommend that the government impose a ban on the use of all harmful ingredients ..."

We must respond to what people want to use on their bodies, and I strongly recommend that the government impose a ban on the use of all harmful ingredients in the manufacture of cosmetics.

The discussion continues on pages 20–21.

A Cosmetic Chemist

My research into the presence of toxic chemicals in cosmetics shows there are significantly higher levels of toxins in household products and even in the environment we live in. These pose far greater health risks to the public than cosmetics do. So, let's keep the subject of this debate in perspective!

"... the cosmetic industry shouldn't have to ban all synthetic chemicals from their products."

There are undeniable advantages to using synthetic ingredients in cosmetics. For example, synthetic ingredients preserve a product for longer. They can also make products easier for consumers to apply. And they are less expensive than natural ingredients, meaning that our products cost less for consumers to buy.

Furthermore, these kinds of cosmetics are only applied in small amounts to small areas of the face. The traces of harmful ingredients like heavy metals are very small. The ongoing research of cosmetic chemists, as well as the use of technology, have enabled even traditional cosmetics to use a much greater proportion of natural ingredients than in previous times.

Under these circumstances, the cosmetic industry shouldn't have to ban all synthetic chemicals from their products.

A Cosmetic Company Chairman

"... we will continue to make cosmetics with some synthetic ingredients ..."

While it is true that established cosmetic companies, like ours, are buying natural cosmetic companies, our loyal customers still want to buy our traditionally produced cosmetics. Our sales haven't declined, and in fact they are steadily growing. Those sales figures send the clear message that our loyal customers buy our cosmetics because they like using them.

Our focus groups and researchers inform us that our loyal customers are not switching to natural cosmetics. Instead, they report that we're attracting new customers by offering a broader range of products that includes natural cosmetics.

All of our products already comply with all existing government guidelines for health and safety. We think that consumers should be able to decide for themselves, and we will continue to make cosmetics with some synthetic ingredients that we believe improve the products.

A Dermatologist Decides

TEXT TYPE
Discussion

It is my professional opinion that more government-initiated tests need to be conducted on cosmetics to identify which ingredients are more harmful than others. That information should be used to update the regulations and manufacturing standards for cosmetic and body-care products.

Cosmetic scientists must also continue to find alternative methods of manufacture that will reduce or eradicate all potentially harmful ingredients from cosmetics, no matter how small their usage.

In the meantime, my best advice for consumers is for you to read the ingredient lists on packaging labels for cosmetic products, and research those ingredients for yourselves. You may discover that some synthetic ingredients, used in minute amounts, are not harmful to people's health.
But you may also decide that too many synthetic ingredients have been used in a product, and that you prefer to use a more natural one. Through education, you can empower yourself to make an informed decision about which cosmetics and body-care products you wish to use.

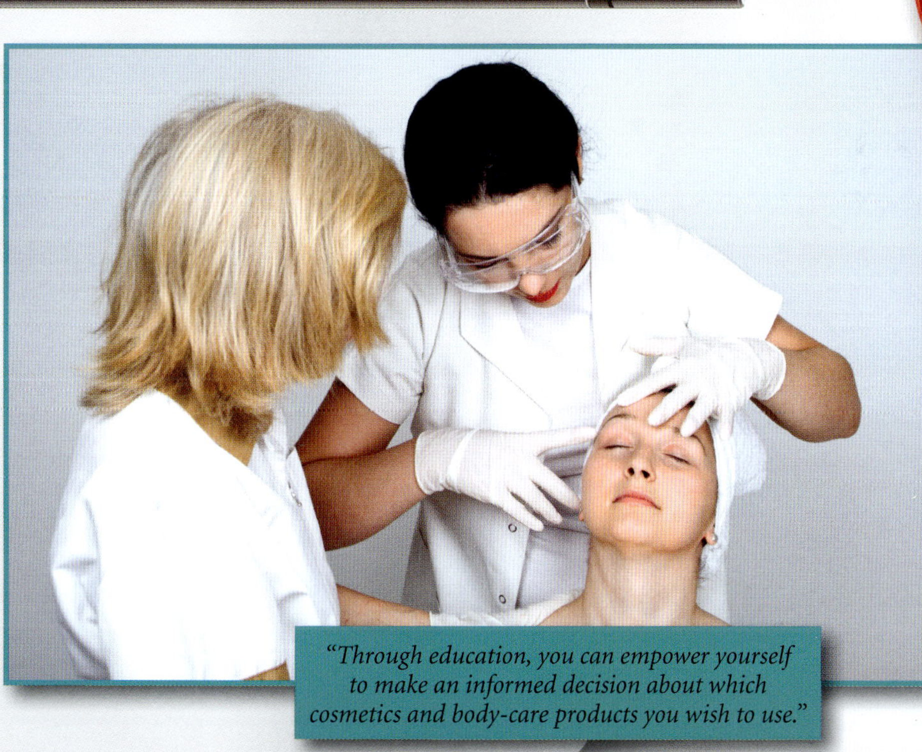

"Through education, you can empower yourself to make an informed decision about which cosmetics and body-care products you wish to use."

The Chemistry of Making Lipstick

The practice of lip colouring has its origins in ancient civilisations. One famous historical figure, Cleopatra VII of Egypt, wore lipstick made from the red colour extracted from crushed carmine beetles. Today, lipsticks account for the largest share of cosmetic sales.

CLEOPATRA VII (69–30 BCE)

Cleopatra VII was Egypt's most famous queen. She inherited the throne at age 18, and ruled 51–30 BCE.

Regulations for Making Lipstick

Cosmetic companies all use a similar method of making lipstick. Since lipstick can be ingested with food when we eat, there are strict government regulations relating to the kinds of raw materials used, as well as the processing, storage and packaging of lipstick.

Main Ingredients in Lipstick

Oils to assist the blending of all the ingredients into a smooth texture
50–70% of the total ingredients

Wax to enable the lipstick to hold its shape. Wax can be used in various forms (for example, powder, solid) and the most commonly used waxes are carnauba, candelilla and beeswax
20–30% of the total ingredients

Pigments to create the desired colour
5–15% of the total ingredients

Other ingredients are used as mixers, preservatives, moisturisers and fragrance. In some lipsticks, ingredients are added for UV protection or to create a pearlescent look.

A professional make-up artist completes the make-up for a character in a medieval performance.

Making Lipstick

Step 1.
The oil is heated in a stainless steel or ceramic vessel.

Step 2.
The waxes are melted in a separate vessel, often called a kettle. Sometimes a little oil is added to assist the process.

Step 3.
The waxes are added to the oil.

Step 4.
The pigments are added to the mixture.

Step 5.
Other ingredients are added to the mixture. These include preservatives, to prolong the life of the lipstick, and fragrances, to mask the scent of the other ingredients and make the lipstick pleasant to smell.

Step 6.
When the lipstick mixture is at the right temperature, it is poured into moulds to cool and solidify.

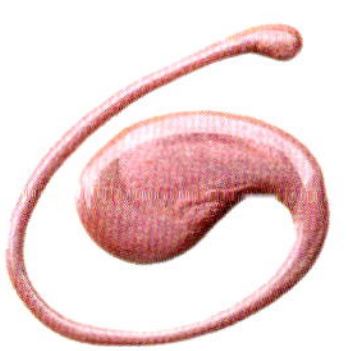

A Lipstick Batch Mixer Speeds Up the Process

A lipstick batch mixer can combine Steps 1–5 (above) into one stage. The advantage for lipstick manufacturers is that all the ingredients can be placed into a high-speed batch mixer at once. The high-speed rotation technology:

- reduces the production times
- improves the texture of the lipstick
- improves the consistency of the mixture.

10 The Chemistry of Toothpaste

Toothpastes can be produced as a white paste or in a gel form. While the formulations for toothpaste and tooth gels are different, they do share a similar combination of ingredients and method of production.

What's in Toothpaste?

Ingredient Categories	Purpose of Ingredients
Liquid (water with some additives)	to prevent toothpaste from drying out
Binders	to stabilise toothpaste and keep the ingredients from separating
Active Ingredient (usually fluoride)	to protect teeth from decay
Sweetener	to counteract some bitter ingredients
Preservative	to preserve toothpaste so that it can be used longer
Abrasives	to create the cleaning and polishing effect
Flavour (usually mint)	to give toothpaste a pleasant flavour
Colouring	gives toothpaste its colour, which is often white or blue
Detergent	to make the toothpaste foam

Toothbrush Care

Dental associations recommend replacing toothbrushes every 3–4 months, or earlier if the bristles have begun to fray. Toothbrushes are best rinsed clean of all food and then stored upright in a clean cup or vessel to air-dry the bristles. It is advisable not to store wet toothbrushes in a closed container as this moist environment may encourage the growth of harmful bacteria.

It's also a good idea is to keep each family member's toothbrush separate, to keep cold and flu viruses from being passed between brushes.

And … never share a toothbrush!

Strict Hygiene Conditions

Since toothpaste is used in the mouth, the ingredients must meet high purity standards and it must be produced under strictly hygienic conditions.

A Method for Making Toothpaste

First, the liquid base is prepared by mixing the additives with the water.

Then, binders are added to the liquid base.

The fluoride, sweetener and preservative are added.

Next, the abrasive ingredients and fillers are blended with the mixture.

Flavour and colouring are added.

Finally, the detergent is added slowly, to keep it from foaming up.

11 A Natural Skincare Company

Sisters Start Skincare Company

trilogy

Sisters Catherine de Groot and Sarah Gibbs grew up in Australia on an avocado farm. After their school years, Catherine became a beauty journalist and Sarah worked overseas before returning to New Zealand, where she set up a company manufacturing plant ingredients for the natural cosmetics industry.

Sarah Gibbs (left) and Catherine de Groot

In 2012, Trilogy's profits from the sale of Helping Hand Wash (above) were donated to the Child's-i-Foundation, *a charity that provides support for abandoned children in Uganda, Africa.*

Rosehip Discovery

While working with plant ingredients, the sisters discovered the natural skin-healing properties of rosehip oil. In 2002, they decided to combine their skills and experience, and formed their own skincare company, Trilogy, using certified organic rosehip oil in their products.

Rosehip-Based Products

The sisters set about discovering exactly what people wanted from natural skincare. After several months of trials and research, they launched the first range of five rosehip oil products for skincare.

Today, the Trilogy range consists of over 40 award-winning products, and the brand has an international reputation for producing ethical, sustainable, high-performance natural skincare.

a handful of rosehip berries

Trilogy's Certified Organic Rosehip Oil

Life Science

Why Rosehips Are Used in Skincare Products

Rosehip berries remain after roses have finished flowering, then oil is extracted from the seeds of the berries. Rosehip oil has healing properties for the relief of skin problems, such as sunburn, eczema, rashes and scarring. Some skincare products use rosehip oil as its main ingredient, as it has proven to rejuvenate and heal the skin for many people.

Açaí Berries in Skincare Products

The açaí berry is harvested from a species of palm tree, the açaí palm, grown in Central and South America. Trilogy uses açaí berry seed oil in some of their skincare products because it is rich in antioxidants, which help keep the skin healthy and looking youthful. When consumed, the açaí berries assist the digestion of food and the flushing out of toxins from the body.

Trilogy's **Skincare** Product Cycle

A Trilogy cosmetic chemist explains, "The skincare products are handmade by a team of dedicated staff – with the help of machines – but they are not produced on an automated production line."

1. MARKET RESEARCH

Making a new natural skincare product all starts with market research into current trends and what the next big thing might be – the whole team gets involved in this part, suggesting ideas and talking about what we could make that will help us stand out among our competitors.

2. PRODUCT AND FORMULATION BRIEF

Our research, development and formulations specialist creates a product and formulation brief for our cosmetic chemist, the person who actually makes the product, covering:

- what the new product is
- what we want it to do and how we want it to perform
- how we want it to feel when someone uses it
- how we want it to smell
- who we think will want to buy it.

3. RESEARCH RECIPE AND CREATE THE PROTOTYPE

The cosmetic chemist and the formulations specialist spend a lot of time researching ingredients and different ways to formulate the product. Once they think they've got the recipe just right, the chemist creates a prototype formulation. This part of the process usually takes around three months.

sacks of rosehip berries ready for cold-pressing

açaí berries are also used in Trilogy products

4. TEST PROTOTYPE SAMPLES

The new prototype product samples are then tested in eight-week cycles with human volunteers – often members of the Trilogy team – to see how different people like it and how it performs on different skin types.

5. FORMULATION IS PRE-APPROVED

If the product meets with all the criteria in the brief and people enjoy using it, the formulation is pre-approved. Sometimes we need to make tweaks to the recipe, and if this is the case, we go back to step three and run through the process again.

"Yes, my skin feels smoother and healthier."

6. UPSCALE BATCH TESTING

The next stage is called upscale batch testing – we make a larger quantity using the same recipe and test it to ensure it looks, feels, smells and performs the same as the small prototype batch. This normally takes around ten days, unless it needs more tweaks to the recipe.

7. FORMULATION IS APPROVED

Once the upscale batch testing is finished and we've got the recipe right to make it in large quantities, the final formulation is approved.

filling bottles with formulation in the Trilogy manufacturing facility

8. LABORATORY PRODUCTION

The approved formulation then goes into production at the laboratory.

9. PACKAGING PRODUCTS

The bottled products are packed into Trilogy boxes for distribution and sale.

sorting and packing Trilogy products

10. MARKETING PRODUCTS

Trilogy informs consumers about the benefits of their products via many marketing activities.

a marketing product launch of new Trilogy products

12 Adolescent Acne

Acne is a skin condition that affects most of us at some stage of our lives. The onset of acne most commonly occurs when the body is releasing more male hormones (testosterone) than normal. This surge of hormones usually lasts until the late teens and will affect every adolescent differently, as some teens may get more acne than others. Although adolescent acne is more prevalent during the teenage years, it can start from the age of ten.

Acne can surface as pimples, blackheads or whiteheads. Many over-the-counter acne treatments are designed to clear the acne, but if the problem persists it is advisable to see your family doctor or a dermatologist.

A hair follicle is the tiny sac or area from where a hair grows and into which oil is secreted from the body's sebaceous glands.

A DERMATOLOGIST

A dermatologist is a doctor who specialises in the diagnosis and treatment of skin conditions.

How Does Acne Form?

The hair follicles (or pores) in skin contain sebaceous glands, which produce sebum, an oil designed to moisturise the hair and skin. Just prior to and during adolescence, hormones in the body cause these glands to produce more oil. When too much oil is produced the skin's pores can become clogged up. The skin's natural bacteria feed off the excess oil and can multiply to such an extent that the tissues around a pore become infected. This causes swelling and inflammation, which shows up on the skin as a pimple.

"Can you get rid of my acne by Saturday night?"

"Hey, it's okay ... most of our friends have acne, too."

Acne Advice

Generally, dermatologists recommend that you wash your face with a mild soap morning and night, to prevent or minimise the effect of acne. By adopting this kind of skincare routine, the skin is kept clean and the amount of bacteria forming on the surface of the skin is reduced.

Do

If you use an acne treatment:

- follow the instructions on the product package
- apply the acne treatment to the pimple as well as the surrounding area

- be patient, as dermatologists advise that an acne treatment can take weeks to work
- once the acne clears, use oil-free products on your skin. (Oil-based products can clog the skin's pores and may cause further acne breakouts.)

Do Not

No matter how hard it is not to squeeze a pimple it is recommended that you do not, because the skin can become irritated. If that happens, a further breakout of acne is more likely to occur.

Don't use acne products that are abrasive or harsh as they can pierce the skin and cause inflammation and infection.

Chemistry of Acne Treatments

Many skincare companies continue to investigate ways to improve the effectiveness of their treatments for adolescent acne. In recent years, cosmetic scientists have discovered acne treatments that are less abrasive and more effective in treating acne.

How Acne Treatments Work

Acne treatments used in the early stages of acne may help to prevent the acne from spreading. Acne treatments that are applied to the skin will not only treat the inflammation but the bacterial infection, too. They kill the bacteria and reduce the secretion of oil from the sebaceous glands in the hair follicle, which in turn reduces the severity of the acne. These combination treatments speed up the healing process and often result in clear skin with minimal scarring.

It's also possible to suffer from acne when your skin is too dry, although it's less common. This is because when skin becomes too dry, the sebaceous glands produce more oil to compensate, and this can clog the pores. Rather than using a standard acne treatment, people with dry skin get better results by using moisturising lotions or creams (as long as they are oil-free). Using a moisturiser stops the sebaceous glands from producing the extra oil, because it's no longer necessary.

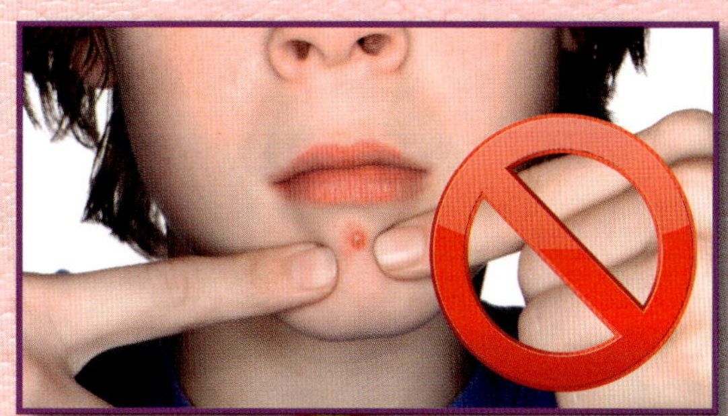

Index

Glossary

abrasive A substance or product that is harsh or rough

accumulate To gather or pile up

airborne Carried by or through the air

consumer A person who acquires goods and services for his or her own personal needs

insatiable A very strong hunger or desire for a product or for knowledge

regulatory To control or direct something according to a particular law

respiratory Relating to the process of breathing, or inhaling and exhaling

rouge A red or pink cosmetic for colouring the cheeks or lips

stipulate To specify or promise something in an agreement

synthetic Not natural or genuine, but made artificially